Eco-Journey

EXPLORING
LAKESHORES

Barbara J. Behm
Veronica Bonar

Gareth Stevens Publishing
MILWAUKEE

For a free color catalog describing Gareth Stevens' list of high-quality books, call 1-800-341-3569 (USA) or 1-800-461-9120 (Canada).

ISBN 0-8368-1065-1

North American edition first published in 1994 by
Gareth Stevens Publishing
1555 North RiverCenter Drive, Suite 201
Milwaukee, WI 53212, USA

This edition © 1994 by Zoë Books Limited. First produced as *Take a Square of Lakeshore* © 1992 by Zoë Books Limited, original text © 1992 by Veronica Bonar. Additional end matter © 1994 by Gareth Stevens, Inc. Published in the USA by arrangement with Zoë Books Limited, Winchester, England.

Photographic acknowledgments
The publishers wish to acknowledge, with thanks, the following photographic sources:
t = top *b* = bottom
Cover: Helen Crabtree; Title page: Stephen Dalton/NHPA; pp. 6*t* David Woodfall/NHPA; 6*b* Hans Reinhard/Bruce Coleman Ltd.; 7 Jane Burton/Bruce Coleman Ltd.; 8, 9*t* Kim Taylor/ Bruce Coleman Ltd.; 9*b* Gordon Langsbury/Bruce Coleman Ltd.; 10*t* Stephen Dalton/NHPA; 10*b* Hans Reinhard/Bruce Coleman Ltd.; 11 John Shaw/Bruce Coleman Ltd.; 12 Jane Burton/ Bruce Coleman Ltd.; 13*t* Lutra/NHPA; 13*b*, 14, 15*t* George Bernard/NHPA; 15*b* Jane Burton/ Bruce Coleman Ltd.; 16*t* Dr. F. Sauer/Bruce Coleman Ltd.; 16*b*, 17, 18 Jane Burton/Bruce Coleman Ltd.; 19*t* Kim Taylor/Bruce Coleman Ltd.; 19*b* George Bernard/NHPA; 20, 21*t* Jane Burton/Bruce Coleman Ltd.; 21*b* Joe Blossom/NHPA; 22 John Shaw/Bruce Coleman Ltd.; 23*t* Konrad Wothe/Bruce Coleman Ltd.; 23*b* Stephen Krasemann/NHPA; 24*t* B. & C. Calhoun/ Bruce Coleman Ltd.; 24*b* Hellio van Ingen/NHPA; 25 Stephen Dalton/NHPA; 26*t* Laurie Campbell/NHPA; 26*b* Sorenson & Olson/NHPA; 27 Stephen Krasemann/NHPA.

Printed in the United States of America

1 2 3 4 5 6 7 8 9 99 98 97 96 95 94

Title page:
A frog uses its strong back legs
to leap away from danger.

Contents

Words that appear in the glossary are printed in **boldface** type the first time they occur in the text.

The lakeshore in spring

In spring, insects that have spent the winter buried in the mud of the lakeshore come out to eat. They eat tiny plants called **algae**.

▲ A mist lies low over the lake. A swan glides past tall reeds at the lakeshore.

▶ Yellow marsh marigolds and tall blue irises grow at the lakeshore. White forget-me-nots flower in the shallow water.

◄ Toads spend winter on land – in holes or under logs. In the spring, the female lays eggs in the form of long ropes that wrap around stems of water plants.

At the lakeshore, frogs and toads catch insects. Swans and ducks eat seeds, small fish, and plants. Voles nibble water plants, and water shrews hunt for worms and shrimp.

Mating and nesting

The lakeshore is home to a fish called the stickleback. In spring, the female lays the eggs, and the male **fertilizes** them.

▶ When stickleback eggs hatch, the male looks after the young ones.

8

◄ Most adult mayflies live for only a few hours. The male dies soon after mating. The female dies soon after she has laid her eggs.

▼ Birds called reed warblers live at the lakeshore. They feed their babies mayflies and other insects.

Mayflies hover above the lake in a rising and falling dance. The females lay eggs that hatch into **nymphs**. In time, the nymphs grow into adults.

9

On the water

Plants called pondweed provide a resting place for certain insects. The insects live under water but come to the surface to breathe.

▲ Some insects, such as the pond skater, live on the surface of the water. Hairs on the pond skater's legs help it "skate" over the surface.

▶ The long stems and flower spikes of pondweed are held up by the water.

The broad leaves of pond-weed also provide shelter for insects and fish. Water birds, such as swans, coots, and gallinules, feed on the leaves, seeds, and roots of the plants.

▲ Gallinules swim with a bobbing action. Their long, thin toes help them walk over the soft mud of the lakeshore.

Under the water

Many water plants grow almost completely under the water. Some have roots in the mud. Others, such as hornwort, float freely in the water.

► A pair of crested newts searches for food among the hornwort plants. Newts are salamanders that live on land for most of the year. In spring, they enter the water to mate. As the female newt lays each egg, she attaches it to a leaf of pondweed.

◄ Freshwater shrimp feed on dead or decaying leaves that have fallen into the water. Shrimp usually live under stones or on the soft mud.

▼ Snails feed on algae and other plants. They scrape off bits of the plants with their rough tongues.

Shrimp and crayfish are **crustaceans**. They are able to breathe under the water through **gills**. Some water snails also have gills, but others must go to the surface for air.

Food for all

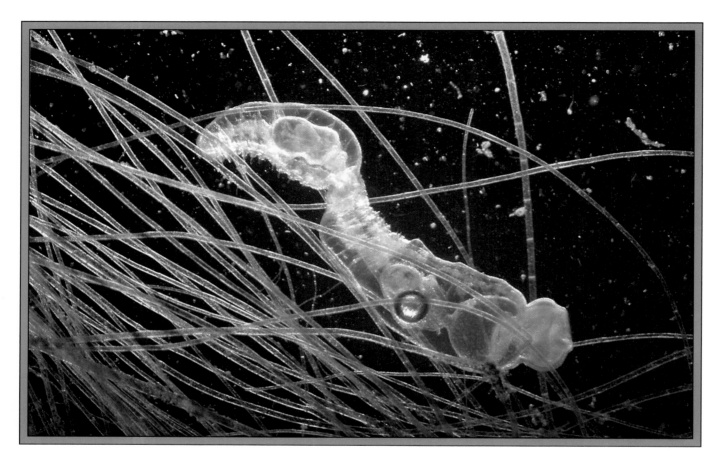

▲ Some worms have colorless, see-through bodies. They are often found in the mud, joined together in long chains.

Insect **larvae** and worms burrow in the lakeshore mud and feed on decaying algae and other plants they find there.

Larvae and worms in the water are eaten by other insect larvae, frogs, fish, and birds. The caddis fly larva protects itself from **predators** by building a tube around its body made of leaf pieces or pebbles.

▲ A leaf leech has a small sucker at its head and a large sucker at the end of its body. A leech holds onto its victim with the large sucker. It uses the smaller one to suck a victim's blood.

◄ Algae plants grow in long chains. They are held up in the water by bubbles of oxygen that the plants make.

Hunters and hunted

Small animals, such as amoeba and hydra, eat **plankton**. Plankton are so tiny that a microscope is needed to see them.

▲ Hydra attach themselves to plant stems by a sticky fluid in their tails. Some hydra inject a liquid into their prey to paralyze them.

▶ This diving beetle has caught a young stickleback to eat. The diving beetle also eats water insects, tadpoles, and small frogs.

16

◀ This water spider has caught the nymph of a dragonfly. The spider will take its prey back to its home to eat.

The water spider uses air bubbles to make a bell-shaped home. The spider leaves the bell only to find food. It spins a web to catch its prey.

Summer insects

A dragonfly nymph may spend its first one to five years under water. Then its skin splits open, and the adult dragonfly flies off.

▶ This dragonfly has just come out of its skin. The skin hangs on the leaf below.

◀ Damselflies rest with their wings folded over their backs, as pictured. Dragonflies rest with their wings spread out.

▼ The midges caught in this spider's web will make a good meal for the spider.

On summer evenings, birds skim over the calm waters of the lake, eating insects called midges. Spiders also catch the midges in webs they have built on the lakeshore.

Freshwater fish

Fish, such as carp and perch, lay eggs near the lakeshore. Trout feed on insect larvae, beetles, and small crustaceans there.

▶ During the day, crayfish (pictured at bottom) hide underneath large stones, safe from predators. Crayfish search for food, such as insects, frogs, and small fish, at night.

20

◄ At spawning time, the belly and fins of male minnows turn orange or red.

In early summer, minnows gather in large groups called shoals. They spawn, or lay their eggs, in gravel patches on the lakeshore. They feed on midge larvae.

▼ After six years in a lake, young elvers become adult eels. They change color from yellow to silver and swim downstream to the sea.

Waders and water birds

Herons, grebes, bitterns, and kingfishers are all lakeshore birds. They nest at the lakeshore, where food is plentiful.

▶ A long-legged heron waits at the water's edge. It uses its sharp, pointed beak to spear fish and frogs in the shallow water.

◀ Birds called dippers can dive or walk under the water. They live near fast-running water, where a stream flows out of a lake.

▼ Bitterns make their nests in the reeds on the lakeshore. If they are frightened by an enemy, they stretch up their necks and bills to look like the reeds.

Many water birds spend a long time preening, or cleaning, their feathers. Preening removes dirt and insects and spreads oil over the feathers. The oil makes the bird's feathers waterproof and smooth.

Lakeshore animals

Otters are the largest of the lakeshore animals. They spend the day in a burrow, or hole, in the bank of a lake. At night, they hunt fish, frogs, eels, and birds that they eat on the shore.

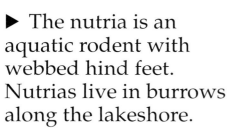

▲ Otters are playful animals that often slide on the ice in winter.

► The nutria is an aquatic rodent with webbed hind feet. Nutrias live in burrows along the lakeshore.

24

◄ The water shrew is only about 3.5 inches (9 centimeters) long. It has a long, pointed snout plus hairs on its paws that help it swim.

Water voles have very thick fur. They feed on water plants. Tiny water shrews eat insects, worms, tadpoles, and small fish. They catch their prey with very sharp teeth.

▶ In early winter, the female brown trout chooses a patch of gravel near the place where a stream enters a lake. The moving water keeps the gravel clean. She digs her nest there and lays her eggs. The male fertilizes the eggs, which then sink into the gravel.

▼ Winter is hard for water birds like these mallard ducks. When the lake freezes over, they cannot reach the food they need beneath the ice.

The lakeshore in winter

As winter comes, many birds migrate, or travel long distances, to warmer places. Fish swim very slowly or rest on the lake bottom. They do not use much energy, and their body temperature is low.

In winter, the water at the bottom of a lake stays above freezing. Because of this, many water animals can survive the winter.

▼ Snow geese spend the summer in breeding grounds far north in the Arctic. They spend the winter at the lakeshore.

More Books to Read

The Bird Book. Neil and Karen Dawe (Somerville House Book, Workman Publishing)

Discovering Worms. Jennifer Coldry (Bookwright Press)

Frogs, See How They Grow. D. Kindersley (Lodestar Books, Dutton)

Plenty of Fish (a *Science I Can Read* book). Millicent Selsam (Harper and Row)

Pond and River. Steve Parker (Knopf)

Snail. Jens Olesen (Silver Burdett)

Videotapes

Call or visit your local library to see if these videotapes are available for your viewing.

The Frog. Animal Family Series (Barr Films)

Little Tadpole Who Grew Up. Coronet (MTI Film and Video)

Water. Primary Science Series (Barr Films)

28

Places to Write

For information regarding nature centers in your area, contact:
National Audubon Society
700 Broadway
New York, NY 10003

For more information regarding lakes and wildlife, contact:

United States Department
  of the Interior
Fish and Wildlife Service
Publications Unit
Washington, D.C. 20240

Internal Ministry of the
  Environment
Public Information Center
First Floor
135 St. Clair Avenue, West
Toronto, Ontario M4V 1P5

Interesting Facts

1. Fish live in every kind of water except in very polluted water or very salty water such as the Great Salt Lake of Utah in the United States and the Dead Sea between Israel and Jordan.

2. Fish use their fins to balance, steer, and stop in the water. They wiggle their bodies and tails to swim. They can also shoot a stream of water out of their gills to push off from a resting position.

3. Trout are fast swimmers. They can swim up to 23 miles (37 kilometers) per hour.

4. The male stickleback cares for the eggs by fanning them with his fins to give them fresh water.

5. A water snail can hang onto the underside of the surface of the water with its wide "foot."

6. Water scorpions have a tube at the end of their bodies. They push the tube up above the water to breathe.

7. Female damselflies use their tails to cut holes inside the stems of underwater plants. They lay their eggs inside the plant stems.

8. After swimming downstream out of a lake, eels migrate all the way across the Atlantic Ocean to the Sargasso Sea before they lay their eggs.

9. Water birds use oil from the preen gland at the base of their tails to waterproof their feathers.

10. Herons have special feathers on their chests which they nibble to break up into a powder. They use the powder to clean all their feathers.

Glossary

algae: tiny plants that live mostly in water.

crustacean: an animal with a crust or shell over its body for protection.

fertilizes: joins male cells and female cells together so a new plant or animal can grow.

gills: the parts of fish, snails, or crustaceans that are used for breathing. The gills take oxygen out of water.

larvae: the second stage in the life of certain insects, after the eggs have hatched.

nymphs: certain insects, such as the mayfly, dragonfly, and damselfly, that are at the stage between an egg and an adult.

plankton: tiny animals and plants, such as algae, that float in the water.

predators: animals that kill other animals for food.

Index